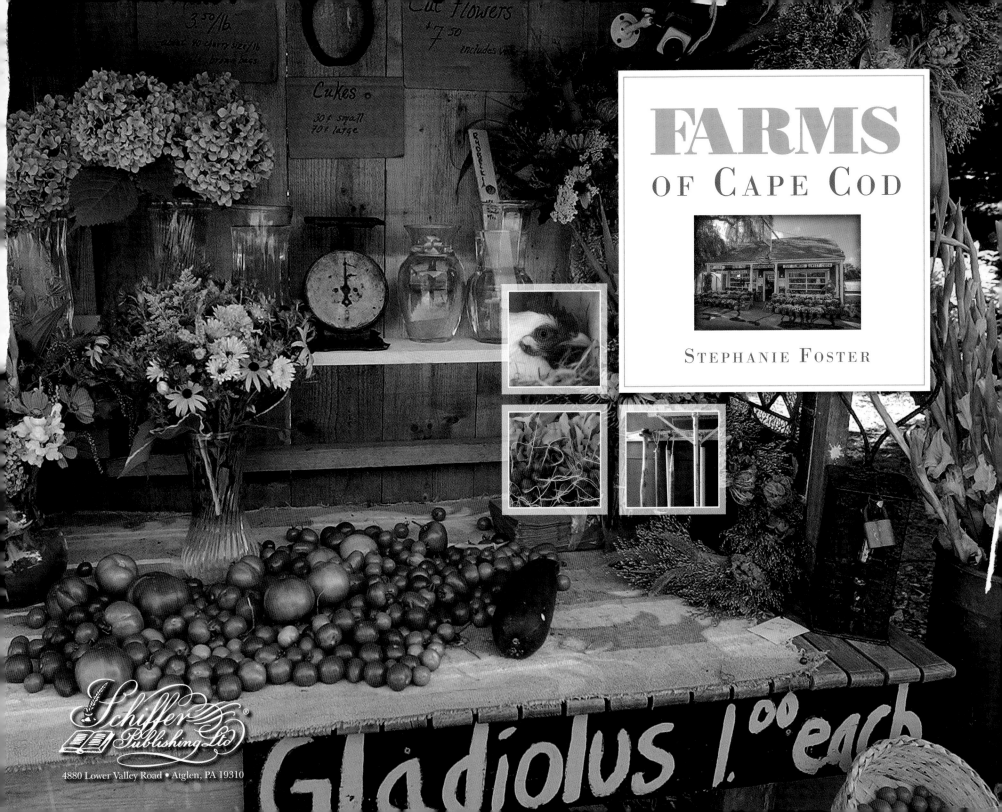

FARMS
OF CAPE COD

STEPHANIE FOSTER

Schiffer Publishing Ltd

4880 Lower Valley Road • Atglen, PA 19310

Library of Congress Control Number: 2013932308

Designed by Danielle D. Farmer
Cover Design by Justin Watkinson
Type set in Bodoni Poster/Bodoni MT/Edelsans/Helvetica Neue LT Pro

ISBN: 978-0-7643-4432-9
Printed in China

Published by Schiffer Publishing, Ltd.
4880 Lower Valley Road
Atglen, PA 19310
Phone: (610) 593-1777; Fax: (610) 593-2002
E-mail: Info@schifferbooks.com

For our complete selection of fine books on this and related subjects, please visit our website at www.schifferbooks.com.

You may also write for a free catalog.
This book may be purchased from the publisher.
Please try your bookstore first.

We are always looking for people to write books on new and related subjects. If you have an idea for a book, please contact us at proposals@schifferbooks.com.

Schiffer Publishing's titles are available at special discounts for bulk purchases for sales promotions or premiums. Special editions, including personalized covers, corporate imprints, and excerpts can be created in large quantities for special needs.
For more information, contact the publisher.

In Europe, Schiffer books are distributed by
Bushwood Books
6 Marksbury Ave.
Kew Gardens
Surrey TW9 4JF England
Phone: 44 (0) 20 8392 8585; Fax: 44 (0) 20 8392 9876
E-mail: info@bushwoodbooks.co.uk
Website: www.bushwoodbooks.co.uk

This book is dedicated to

JEAN IVERSEN,

a pioneer in organic farming
on Cape Cod and to all who
follow in her footsteps.

*You can make a
small fortune in farming —
provided you start with
a large one.*

—Unattributed

ACKNOWLEDGMENTS

I want to thank my husband Frank Foster for all his help and support, and Helen Addison of Addison Art Gallery who encouraged me every step of the way.

FOREWORD

With the current proliferation of farmers markets, farm stands, and CSAs, it's astonishing to recall that less than ten years ago there was only one farmers market on Cape Cod where you could purchase produce directly from the grower. Even more amazing is that you can now also purchase locally raised poultry, pork, lamb, and beef on this peninsula renown for its seafood. While there has been a sizable increase in the number of farms across the state in the past decade—the first increase since WWII—these businesses will only thrive as long as consumer demand for locally grown food continues to grow. So hopefully this is not a passing fad like the back-to-the-earth movement of the 1970s, but a full-blown transformation in the way we purchase our food. Because despite all the arguments for why we should be eating locally grown food, the best reason is because it really does taste better.

Dianne Langeland,
Co-publisher, *Edible Cape Cod*

CONTENTS

INTRODUCTION

Farming is not an easy way of life. It's risky, weather dependent, and labor intensive. But few people experience the level of satisfaction that comes with farming. Some farms have been handed down through the generations like Crow Farm in Sandwich and Kane Farm in Truro. Others, like Halcyon in Brewster, are newly born, thanks to owners who want to grow organic food and share it with their neighbors.

There are vineyards, herb farms, tree farms, cranberry bogs, oyster farms, horse farms, historic farms, and backyard farms scattered across the Cape. All help to keep it green. They survive because of the support and enthusiasm of local customers at farm stands and markets. Without this, farms would become a memory of the past.

When farmland is built on, it's lost forever and that loss would diminish life on the Cape for all of us. This book celebrates both small and large farms and the people who dedicate themselves to them, as well as the tiny stands and carts that dot the side roads offering fresh produce, eggs, and flowers to those who pass by.

CAPE COD
ORGANIC
FARM
CAPE COD

Tim Friary and daughter Emilie of Cape Cod Organic Farm get playful when a litter of pigs start chewing their shoelaces.

Barnstable

The love and knowledge Tim Friary has for farming has grown over the years, along with his acreage. But he is still growing the way his grandparents taught him. Organically. He started out in 1989, while raising his three children, Travis, Sara and Emilie, who worked with him on 14 leased acres. Nineteen years later, he was granted a lease on the Barnstable County Farm.

The historic 63-acre property had provided food for the jail and house of correction for 60 years. Prisoners worked the land growing produce, tending 2,500 chickens and 18 milk cows. Unfortunately, the harvest wasn't reliable and it cost more to grow the eggs than to buy them. When the house of correction moved away, transporting prisoners became unfeasible. The land lay fallow for four years. Then, Friary was granted permission to farm it.

It took several years to refurbish the crumbling infrastructure and restore 30 acres to plantable conditions. Now it's the largest tract of land under cultivation on the Cape. When roosters crow, pigs squeal, and workers till the soil on a picture perfect day, it looks and feels like yesteryear.

The farm is well-known for its variety, quality, and abundance of produce at many of the local farm markets. Customers wait for the market to open to be first in line for whatever is special that week, from strawberries in June to fingerling potatoes in August. Tim's own personal favorite vegetable is brussels sprouts.

The farm stand is open year round and offers organic eggs and whatever vegetables, fruit, and flowers that are in season.

Rosie, the farm dog, seeks a shady spot under a table to rest.

Top left: It took several years for Friary to refurbish the crumbling infrastructure.

Bottom left: Several of the fields are dedicated to growing flowers for Whole Foods Markets off Cape.

The farm offers the largest Consumer Supported Agriculture subscriptions (CSA) on the Cape. CSA customers receive a bag of seasonal vegetables each week, along with a bouquet of flowers and a dozen eggs when the chickens are laying. People can also work in exchange for food.

Friary supplies local gourmet restaurants and upscale groceries with his goods and grows for Whole Foods Market in Massachusetts. He has been able to expand his operation thanks to the public's awareness of the benefits of local, organic, health-giving food. He admits it's a hard life, but says it's a satisfying one. And he never goes hungry.

LEFT TO RIGHT:

Farm workers relax after a long morning in the fields.

Emilie Friary grew up helping her dad both at the farm and the farmers markets.

Julie Davis picks sunflowers for the farm stand on a beautiful summer day.

Emilie named this tiny piglet Gimpy when he was born with an injured limb. He would have died if she hadn't been there at birthing time.

LEFT:
Although the historic property consists of 63 acres, only 30 are tillable.

The farm sells to the public and also offers Consumer Supported
Agriculture memberships.

Tips from the top

Friary starts his brussels sprouts from seed in a greenhouse in late March. Then he pokes holes in black plastic outside and lines them up with irrigation. Around the third week in April he covers them with Agribon cloth (a row cover) to keep insects off until he harvests them in September. His biggest gardening mistake was not using a row cover.

He also discovered that rocks stunt carrots. He advises planting in light soil or loamy sand, not heavy clay. Be sure it's rock-free. Otherwise the carrots will grow crooked and get woody.

Cape Cod Organic Farm, 3675 Main St., Barnstable, is open all year with seasonal organic vegetables, fruit, eggs and nursery stock for sale on their farm stand. A CSA is also offered and produce is brought to local farmers markets from May to November.

FOR MORE INFORMATION

call: 508-362-3675
email: timfriary@aol.com
or check the website at: www.capecodorganicfarm.org.

NOT ENOUGH ACRES
EAST DENNIS

Not Enough Acres is located in a picturesque area,
bordered by marsh and million dollar homes.

East Dennis

Jeff Deck looks every inch the Yankee farmer, although he wasn't born to the soil the way his wife, Beth, was. Growing up, he helped his mother with her small vegetable garden in New Jersey. But he has always been a farmer at heart. Their farm, in a picturesque part of Dennis, has been in the Crowell family since the King's Division and farmed for four generations.

Jeff was a landscaper for 17 years before he switched to serious farming. With only 4.5 acres, he called his farm Not Enough Acres. He eventually remedied the situation by leasing another 3.2 acres of land from his in-laws across the street. He finally has enough acres, but the name stands.

Jeff loves the beauty that surrounds him as much as the people who drive by and take photographs. His chickens roam freely in the salt marsh and his sheep peacefully nibble meadow grass not far from million dollar estates. Uncertainty clouds his future, as his in-laws may be forced to sell their land one day. But Jeff will ride with the punches.

Meanwhile life is good and the labor is pleasant for someone who likes to plant, pick, cook, and eat his own food. He has an eye for color and texture, but only grows what he likes to eat. But he is also adventurous, experimenting with greens of all sorts, from Asian to arugula, along with asparagus, swiss chard, spinach, peas, peppers, potatoes, pumpkins, beets, broccoli, cabbage, carrots, celery, chili peppers, cucumbers, eggplant, tomatoes, herbs, green beans, kohlrabi, kale, leeks, mushrooms, onions, parsnips,

Jeff Deck enjoys the land, growing organic produce and being his own boss at his farm, Not Enough Acres.

Molly hopes it's breakfast time.

NOT ENOUGH ACRES

TOP TO BOTTOM:

Although Jeff Deck now has enough acres it's too late to change the name of his farm.

A brooding hen spends the day on her nest.

Barn swallows holler for mom in the rafters of the shed.

shallots, summer and winter squash, and garlic. Lots of garlic. It's his favorite flavor.

In season, a variety of fruits such as strawberries, raspberries, apples, figs, and pears might be found at his farm stand, if he hasn't sold out at the farmers markets he attends. Because of his high quality, a half dozen local gourmet restaurants clamor for his goods.

Jeff particularly likes being his own boss and enjoys the outdoors and the space around him. For him, the lifestyle is not a hardship. Modest and quiet by nature, his work ethic and industry speak for him. He farms because he loves it.

Beth is responsible for the sheep. Spinning and knitting are her passion and she sells fleece, roving, and yarn. The couple also keeps a flock of chickens and tends to seven bee hives. They allow dandelions to grow wild in the lawn just for the bees.

Beth has an administrative job off the farm. It provides necessary health insurance and a steady, dependable income, but she is a farmer at heart as well and they hope to spend their lifetime there. The couple are aware of their stewardship of the land, which has been in continuous use since 1713, and are writing its history.

CLOCKWISE:

The barn cat keeps law and order on the property.

Grazing sheep add to the peacefulness of the landscape.

Deck generously mulches his summer squash crop with hay from the marsh.

Maintaining a farm is a full time job for this small farmer.

COUNTER CLOCKWISE:

To thrive, vegetables need open fields, lots of sun, and fencing to keep out deer.

Both beets and beet greens are tasty when they are picked the same day they are eaten.

Jeff Deck sells his organic produce at the Harwich Farmers Market as well as the Mid-Cape Market in Hyannis, and has developed a following.

Peppers are popular at the summer market.

17

Deck prefers the natural unadorned taste of the organic vegetables he grows.

Tips from the top

Jeff Deck eschews the use of salt. He prefers the natural flavor of the organic food he grows himself without enhancement. He picks most vegetables in the cool of morning, when the sun is low, but says root crops can be picked anytime. His favorite way to cook Swiss chard and spinach is to sauté them in a pan with garlic.

Not Enough Acres, 107 Sesuit Rd., East Dennis, is open sun-up to sun-down with seasonal organic vegetables, herbs, eggs, fleeces, roving, yarn, honey, maple products, dairy, meat, and wood products. There is usually something from his solar-heated greenhouse, hen house, or beehives year round.

FOR MORE INFORMATION

call: 508-737-3446
email: notenoughacres@gmail.com

HALCYON FARM, Brewster

Brewster resident Lucas Dinwiddle was born and raised on the Cape and had no connection to farming until he went to California. Journalism was his future until he got a job at Fairview Gardens in Goleta, California. Working at a small-scale organic farm and education center changed his mind. His inspiration came from Michael Ableman, a proponent of sustainable agriculture and regional food and the founder of the non-profit center.

He fell in love with the beauty of the food, the growing, and the farm stand displays which approached living art. Dinwiddle switched his major to crop science at California Polytechnic State University's College of Agriculture, Food, and Environmental Sciences. After working on the school farm "without risk," as he puts it, and spending ten years away from home, he was ready to farm on his own and returned to the Cape.

A friend offered him land to plant near Rock Harbor in Orleans, so he planted the specialty crops of strawberries and garlic. When his family bought property right on Route 6A as an investment, it came with an acre of land and the perfect place for a farm stand with high visibility. When they offered to rent it to him, he jumped at the opportunity.

Dinwiddle adjusted to the seasonal aspect of farming on the Cape, working in sandy soil and without a tractor.

COUNTER CLOCKWISE:
Lucas Dinwiddle adds to the ambience at Orleans Farmers Market on Saturdays, when he brings his colorful produce and artfully displays it.

Farmer Dinwiddle can step outside his back door and see his produce growing.

Fresh spring peas are the first to pop in the garden and the first to sell out at the market.

The farm stand awaits the morning harvest on this well-kept small farm.

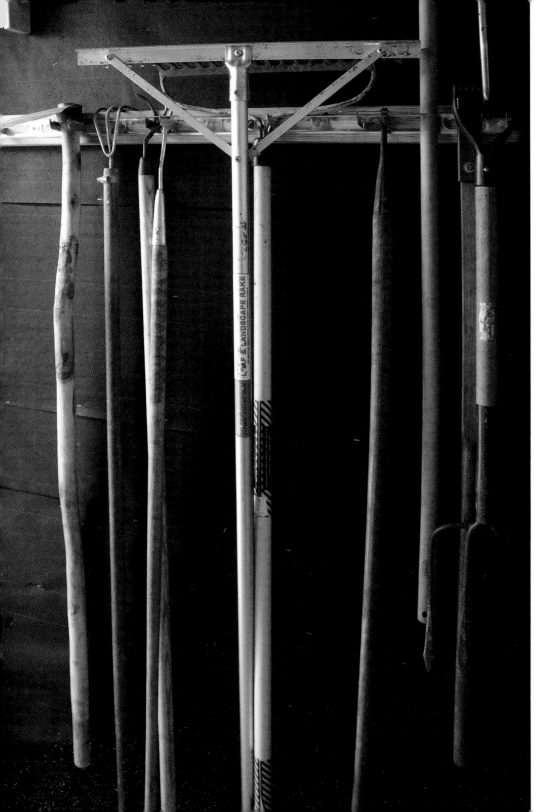

LEFT:
Tools of the trade are neatly arranged in the shed.

But he is a patient man. He put aside thoughts of avocados and tree nuts in his backyard and now his energy goes into slowly enriching the soil.

He is a new breed of farmer who is both thoughtful and artful. A perfectionist, he is in no rush. His CSA is small. He sells to a few gourmet restaurants that appreciate the picture perfect organic produce he provides, like French breakfast radishes, tatsoi, and Russian red kale. His farm stand is the prettiest of all at the farmers markets and his goal is to have a beautiful farm. He loves the idea of small community farms and backyard agriculture. And with all of his youthfulness and enthusiasm, he is optimistic his farm will sustain him.

Halcyon Farm, 3915 Route 6A, Brewster, offers fresh produce and flowers.

FOR MORE INFORMATION

call: 805-540-4103
email: lucasdinwiddle@gmail.com

TOP TO BOTTOM:
Ripe seasonal tomatoes make an arresting display but won't be around for long.

Dinwiddle has adjusted to growing crops on the Cape by amending the soil.

Hemeons Farm, Harwichport

The four-acre farm operated by Brent and Peggy Hemeon became a business when their 7 year-old son, Corey, sold strawberries at the end of the street. The family had always maintained a garden for themselves, but it started getting bigger. Then, Brent became interested in fruit trees. Before they knew it, they had more of everything than they could eat.

Now they have more than 200 fruit trees, including white and yellow peach, two kinds of pears, and a variety of apples. Add in blueberries, blackberries, strawberries, raspberries, watermelon, and cantaloupes, and there is a lot to choose from.

The Hemeons are Harwich natives. Peggy is a Cahoon and her lineage goes back to the Mayflower. They've operated their small farm stand for more than 30 years and, now that the public is interested in buying local, it has become increasingly popular. Tried and true seasonal vegetables can be found on their stand or at the Thursday farmers market in Harwich, although Brent likes to try something new each year.

When their son Glenn became a landscaper, they took over his hydrangea business and propagated plants as well. Peggy admits they were doing all the work anyway. With the addition of two large tunnel greenhouses, there is no end to the crops they can grow. Except there is a limit on time available. As a typical Cape Codder, Brent wears two hats. He is also a contractor.

Each week Peggy Hemeon brings hydrangea plants that her husband Brent propagates to the Harwich Farmers Market, along with seasonal crops grown on their four-acre farm.

Hemeons Farm, 186 off Bank Street, Harwich Port, is open August through Thanksgiving with seasonal vegetables and fruits. Eastham turnips are available for Thanksgiving, as well as squash, pumpkins, and cranberries from local bogs.

FOR MORE INFORMATION

call: 508-432-3947

CLOCKWISE:

Hemeons Farm, down a side road, specializes in fruit and gorgeous hydrangea plants.

A small farm stand with seasonal vegetables awaits visitors at the Hemeon farm.

There are more than 200 fruit trees on the farm with a wide selection of apples.

Nestwood Farm, Truro

Chris Murphy may be a surfer, artist, athlete, chef, and bartender, but he is also a serious farmer and has been for decades. He grows basil, swiss chard, eggplant, hot peppers, and tomatoes for local restaurants on a two-acre plot in Truro, where he also operates a farm stand in season. He shares the space with a couple of other farmers who appreciate the value of locally grown food.

Nestwood Farm, 11 Hatch Road, Truro. Open daily in season; and at the Wellfleet Farmers Market on Wednesdays from 8 to noon.

CLOCKWISE:
Chris Murphy has been a serious farmer for most of his life.

Customers seem to find Murphy whether he is at his Nestwood Farm stand on Hatch Road in Truro or Wellfleet Farmers Market.

The eggplant on his stand looks good enough to eat.

IN THE WEEDS FLOWER FARM,
Brewster

Debbie Demarais, of In the Weeds Flower Farm on Route 6A in Brewster, probably has the prettiest small farm stand on the Cape offering fresh produce, flowers, and seasonal wreaths, from twigs to vines to holiday greens. The stand is freshly stocked each day and is operated on the honor system.

Demarais is a nature lover, Earth Mother, and forager. She comes from a long line of farmers and her knowledge is impressive. Her main crops are tomatoes and hydrangea, but there are many artistic surprises. Go early for the best selection.

In the Weeds Flower Farm roadside stand, 392 Main St., Brewster, is open from early spring, with daffodil bouquets, to late winter, with bittersweet wreaths and holiday greens.

LEFT TO RIGHT:
In the Weeds flower stand on Route 6A in Brewster features flowers, tomatoes, and other vegetables in season.

Flower grower Debra Demarais specializes in bouquets that feature hydrangeas.

Boxwood Gardens, Orleans

Darnell Caffoni is a massage therapist who loves gardening and growing food for market. She and her late husband Rod started Boxwood Gardens in 2000 with a small plot and two 4 x 8 foot raised beds. They decided on raised beds because it was an efficient use of space and allowed them to plant densely and harvest more.

When they started selling produce at the Orleans Farmers Market, they developed a devoted following and increased their production to 39 beds. A niche farm, they are known for 30 varieties of lettuces, 18 varieties of Asian greens, and their spicy mesclun mix. They also grow ten different types of carrots, eight different beets, seven varieties of radishes, and purple string beans. Everything is organic and pesticide free, including the home-made soap made with herbs from the garden.

Since Rod's death, Darnell's mother, Dessie Richards, has become part of the farm. They make a dynamic team, with Darnell dispensing advice on heirloom vegetables and her mother guiding customers in their choice of soap. Painful as her loss was, farming has helped Darnell heal and find peace.

Boxwood Gardens, 27 Captain Doanes Way, Orleans, sells organic produce and soap at the Orleans Farmers Market on Saturdays from 8 to noon, from May to November.

Darnell Caffoni and her mother, Dessie Richards, of Boxwood Gardens, work at Orleans Farmers Market each Saturday in season.

27

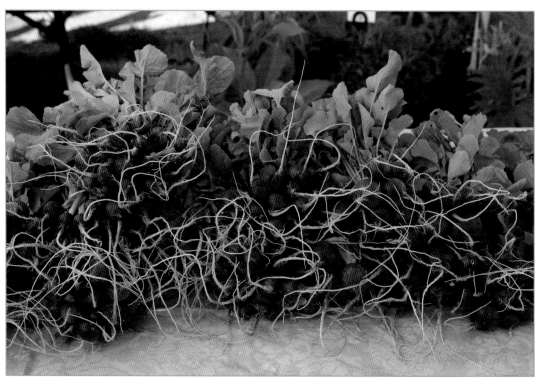

Boxwood Gardens is a niche farm that specializes in unusual greens and root vegetables, rather than compete with larger farms.

A bumper crop of breakfast radishes fills Boxwood Gardens' table at the market.

EAST DENNIS OYSTER FARM
EAST DENNIS

Stephanie and John Lowell of East Dennis Oyster Farm go out to their oyster flats every day of the year, regardless of weather.

John Lowell has been farming oysters in East Dennis since 2003, when few people thought they could make a living at it. With his outgoing personality, he probably should have had his own talk show, but instead he chose to cultivate bivalves. The former landscaper and irrigation specialist left his service business, as did his wife Stephanie, a CPA and justice of the peace. They each applied for and received a one-acre grant in Quivett Neck and oysters have been the focal point of their life ever since.

They started deep in debt but were determined to succeed. The seed oysters only cost two cents apiece, but they needed expensive plumbing. They were raising two children as well, so the first few years were difficult.

It's not a business for an armchair adventurer. The Lowells are out on the sand flats every day. Regardless of the weather. Year round. The tide rules their hours. Hundreds of trays sit above the sand on poles when the tide is out, with gravity holding them in place. Each day they count and grade the inventory according to size and readiness for market.

Oyster farmers must be patient. It takes three years for a seed oyster to develop into an edible commodity. When they reach peak, they are bagged and delivered to local restaurants and fish markets. Lowell is one of 19 farmers in the state with an interstate license and he ships to customers as far away as Texas. He also tags and grows special trays of oysters for weddings or gourmet restaurants. Clients love to drop by and see them. He also shucks his oysters at private parties and restaurants. When his crop is low, he works with other farmers to fill in.

It may sound glamorous, but growing and harvesting oysters is hard work.

LEFT TO RIGHT:
The Lowells find peace and beauty every day at their two acre oyster farm in East Dennis.

John Lowell is a former landscaper and irrigation specialist turned oyster farmer.

Stephanie Lowell gave up her career as a certified public accountant and justice of the peace to work her oyster grant.

In December, the oysters stop growing and harvesting ceases. They put 495 pieces of gear in dry storage and leave five behind. The oysters are stored in a truck, then rotated back to the flats every tide cycle to purge them with cleansing fresh salt water. When March rolls around they are returned to the flats and then sold at maturity. They now have half-a-million oysters in various stages of development on their briny farm.

The Lowells have done a lot to promote their product and the market for it during the last dozen years. Lowell has hobnobbed with the governor and other state officials and shucked oysters at the State House. He's proud of his product and loves to show visitors around his oyster farm. Their outreach has benefited other oyster farmers and all oyster lovers.

LEFT TO RIGHT:
An occasional oyster doesn't make it to market when it's near dinner time.

Oysters are graded by size and put into plastic buckets before being tagged and put back into trays.

It takes patience to grow oysters; in fact, it takes three years before they are an edible size.

The East Dennis Oyster Farm now has now has more than half a million oysters in varying stages of development.

Tips from the top

- Get seed oysters from an approved hatchery.

- Don't start with more than 25,000.

- Be prepared to wait three years as the oysters go through various stages till maturity.

- More seed oysters can be added each year so that the harvest continually rotates.

- Expect possible pests and diseases, just like any other farm.

East Dennis Oyster Farm is located on the tidal flats off Quivett Neck in East Dennis, Massachusetts. Due to the tide schedule, farm tours are by appointment only.

TO ARRANGE FOR AN INFORMATIVE TOUR

call: 508-398-3123
website: www.dennisoysters.com

TOP TO BOTTOM:

The tide rules the hours that the Lowells can go out on the flats.

When you grow oysters, life is a beach.

The Lowells provide gourmet restaurants with oysters and also shuck at private parties.

After three decades, Jean Iversen has her routine down perfectly. First, she harvests her crops, then she stops for her mail, then she goes to her farm stand and unloads her cart.

KELLY
FARM
CUMMAQUID

Since 1981, organic farmer Jean Iversen has added three yards of cow manure, green sand, and rock phosphate to a section of her garden every fall. The result is soil the color of milk chocolate. This nonagenarian founded the Cape Cod Organic Garden Club and is considered a living legend, having mentored many of the best known names in organic farming on the Cape.

Every March, she uses the warmth of her wood stove to germinate her seeds, then moves them to a sunny kitchen window. She repeats the process until she has enough seedlings to fill her fields. One plot, just outside her door, is 140 by 80 feet. Another, down the road, is 50 by 20 feet. A lot of land for a wee lady to work alone.

LEFT TO RIGHT:
Every Tuesday through Saturday in season, Jean Iversen shows up at the Kelly Farm stand on Route 6A in Cummaquid with produce she has grown herself.

A little bit of most everything Iversen is growing appears on her stand, but she saves her favorites for herself.

She starts with the hardiest crops first: broccoli, kale, and onions. Later, her fields are laden with asparagus, leeks, strawberries, peas, soybeans, beans, raspberries, onions, oriental and regular eggplant, potatoes, garlic, carrots, sweet potato, peppers, and her favorite heirloom tomatoes: Rutgers, Celebrity, Brandywine, Sweet 100, and Juliet. Then ambrosia melons and carrots.

She hand picks whatever is in season daily, then carts them to her stand around the corner, stopping briefly to pick up her mail. A vegetarian for a half a century, she has first pick of the fruits of her labor and doesn't put it all on her stand. Most days she sells out by 2 p.m.

Iversen moves quickly. At her age, she doesn't have time to waste. Her doctor told her she was developing arthritis, but she doesn't have any pain. She walks two miles and exercises every day to stay strong. She works without complaint, operating a five-horsepower tiller that can fit between the rows. In the spring she puts down her own drip irrigation, then removes it when it's time to till again the next spring.

Gardening enriches her spiritually and she loves having her hands in the soil. She shares this love by inviting neighborhood children to grow their own plants on her farm. Every time she sees a seed germinate, it reaffirms her belief in God.

CLOCKWISE:
Iversen knows exactly where each item is going before she reaches her farm stand.

Organic onions are grown with love.

Fresh green beans and cukes will go home with the early bird shoppers.

37

Tips from the top

Iversen is an artist when it comes to growing beautiful produce.

A home gardener can plant three to four bean crops, two weeks apart in a four-foot space. The more ambitious, like Iversen, can plant nine crops of beans 10 days apart.

Pick tomatoes, green beans, and squash in the late afternoon and the rest of your vegetables in the early morning.

Save your own seeds. Pick a perfect vegetable like a tomato or pepper that is very ripe and let it sit until it is almost spoiled. Then cut it open and take the seeds out. Clean and let dry on a paper towel.

If you plant two kinds of peppers next to each other, the seeds won't be true.

Kelly Farm Stand, Route 6A and Marston Lane, Cummaquid. Open Tuesday to Saturday, 10 a.m. till sold out, from July to mid-October. Freshly stocked each day with produce, flowers and berries.

FOR MORE INFORMATION

call: Jean Iversen at 508-362-8136

TUCKERNUCK
FARM
WEST DENNIS

In addition to being the Lettuce Queen, Veronica Worthington, owner of Tuckernuck Farm, is a sheep fancier.

Tuckernuck Farm is hard to find for those who have never been there, because people keep stealing the sign. It doesn't faze Veronica Worthington, however. She's been growing produce and flowers, and raising farm animals in the neighborhood for so long, most people know her.

Worthington was born with two green thumbs and grew up with a menagerie of animals, including horses, goats, chickens, ducks, dogs, and a pig. She learned about herbs from her mother and grandmother and lives on the family homestead with her mother, daughter, and granddaughter, making it a four generation household.

Although her interests have changed over time, they are always based on the land. She started out growing flowers and herbs, which she sold in her ancient quahog shed along with herb teas, potpourri, and natural wreaths fashioned from lacy artemisia, larkspur, feathery grasses, and other everlastings.

Then she became interested in unusual plants. She bought exotic seeds and ended up growing rare stock for nurseries. When farmers markets beckoned, she increased her crops of heirloom lettuces and organic vegetables so that she could sell them along with pails of sunflowers.

She built a greenhouse so she could work year round and grow her own plug trays. At the farmers markets, she is such a presence, with her 35 varieties of lettuce in rainbow shades, that she is called the Lettuce Queen. Each head is like a bouquet, and gourmet restaurants and upscale stores beg for her goods.

Worthington grows a vast array of heirloom lettuces for her own farm and for Pleasant Lake Farm.

BOTTOM RIGHT:
The Tuckernuck Farm sign is safer by the shed rather than on the road.

Perennially tanned and barefoot, she not only grows at home on her own farm, but at Pleasant Lake Farm as well. The lady barely sleeps. Now sheep have become her passion and the Shetland variety, her favorite. She raises them for wool, but they are family to her. She loves bringing them to shows and displaying the ribbons they win. And making lovely knitted goods from their fleece.

In her youth, Worthington worked as a boat captain and commercial fisherman and traveled to North Africa, the Bahamas, Key West, and Mexico. Somehow she always managed to garden. It is her passion and her destiny.

LEFT TO RIGHT:
Worthington grows plug trays for herself and other farmers.

Should I stay indoors today or go outside?

Sheep have become a passion for Worthington and they are treated like part of the family.

41

Spring brings a new family member to the barn.

Rain gear, gloves, and a net, because you just never know when a fish might come along.

Tuckernuck Farm, 89 Fisk St., West Dennis, is open May through December. The stand offers heirloom lettuces, assorted vegetables, sunflowers in season, a Salad Club, CSA, raw wool, whipped up wool birds' nests, pipe cleaner sheep, and other novelties.

FOR MORE INFORMATION

call: 508-364-5821
email: veronica_worthington@yahoo.com

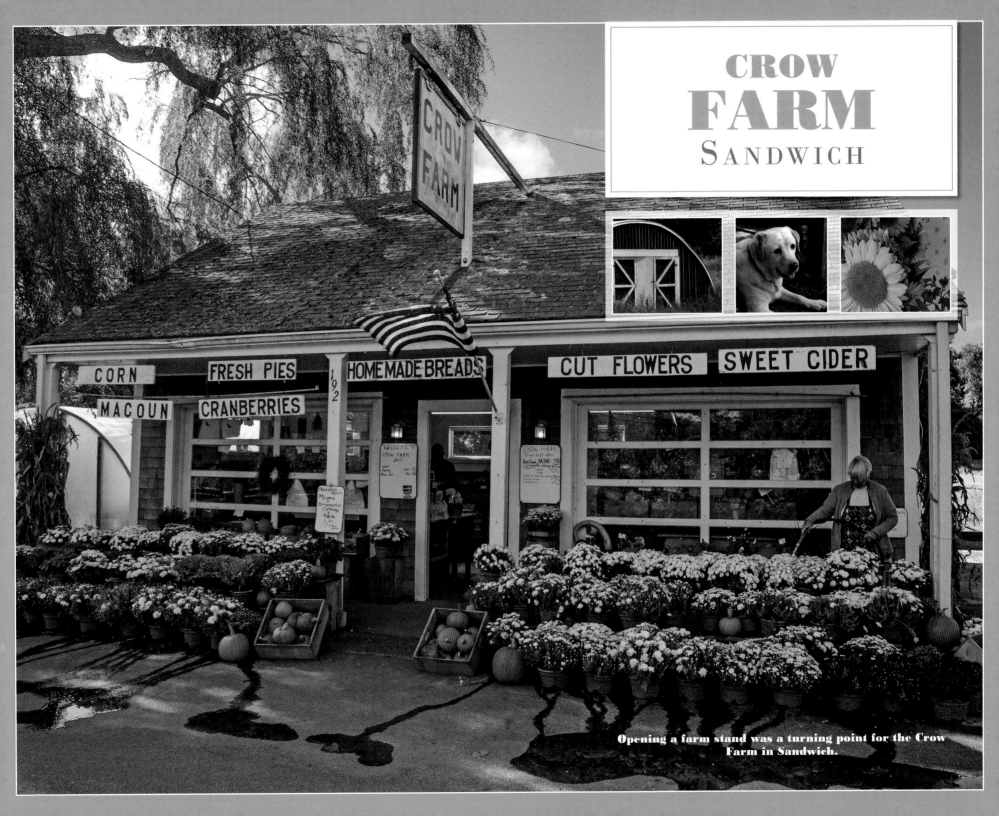

CROW
FARM
SANDWICH

CORN

MACOUN

FRESH PIES

CRANBERRIES

HOMEMADE BREADS

CUT FLOWERS

SWEET CIDER

CROW FARM

192

Opening a farm stand was a turning point for the Crow Farm in Sandwich.

Howard Crowell could have been a millionaire if that was what he wanted. The Sandwich native owns 60 acres of rolling countryside on Route 6A, close to the center of town. One high area even has a view of the sea. Instead of making a developer's dream come true, he farms it, as his father did before him. His son Paul is of the same mind, and now grandson Jason, the fourth generation to farm the land, has made the same choice—to live a sustainable life.

That's why those rolling fields are filled with golden ears of corn and ruby-colored peaches, rather than trophy homes. Sisters and aunts live nearby. All have roots that go way back in the area. David Crowell, Howard's father, had a degree in agriculture from the University of Maine and started the 60-acre farm in 1916. His brother Lincoln, a civil engineer and forester, joined forces with him. Lincoln is known as the person responsible for buying 18,000 acres of land for Camp Edwards, which forms the largest part of the Massachusetts Military Reservation and includes Otis Air National Guard Base and Coast Guard Air Station Cape Cod.

But the brothers couldn't make a living sending produce to Boston. Lincoln went back to being a forester. Later, the nearby Shawme-Crowell State Forest in Sandwich was named after him.

David struggled along until 1930. Suddenly, a new road, Route 6A, brought traffic by his door. It was the start of a retail operation and a life saver. Howard, who was born in 1924, took to farming and worked beside his dad until his death in 1972. He brags he could take any tractor apart and put it back together with his eyes closed. Actually, it's something every serious farmer needs to be able to do.

Howard Crowell, grandson Jason, and son Paul have all made the same lifestyle choice to farm the land that has been in the family for four generations.

TOP TO BOTTOM:

Corn is a big seller at Crow Farm.

Tomatoes are everyone's favorite in the summertime.

45

Forty of the sixty acres are under cultivation and plantings are staggered so the harvest is continuous. They grow bedding plants in 15 greenhouses and they are their number one crop. Two-thirds of the business is wholesale. Professional landscapers come in for their plants, using an old-fashioned honor system. The farm also provides local towns, as well as a hotel chain, with plants.

Corn and tomatoes pull customers into the farm stand, and they are known for fresh peaches. They don't grow celery, beets, potatoes, carrots, or garlic, because they can't make a living at it. They know what they can sell and have survived poor years because of their diversity.

A good irrigation system is one of the secrets to their success. They have improved and upgraded theirs over the years so that it now is state of the art with a mobile system and drip irrigation, both labor-saving devices.

Paul and Jason have expanded the crop list with Macintosh, Macoun, Cortland, Honey Crisp, and Russet apples from their apple orchard. And fresh raspberries. The Crowells like the work because it is varied and changes with the seasons. Their philosophy is simple. "If you don't like what you are doing, you won't do a good job."

Paul and Jason Crowell expanded the crop list with a variety of apples.

LEFT TO RIGHT:
At Crow Farm, they know what will sell and don't bother with the rest.

Bedding plants have become a big wholesale business.

Gomphrena is grown for both fresh and dried bouquets.

The Crowells enjoy the change of seasons and crops.

Tips from the top

Flowers are grown for mixed bouquets that are sold at the farm stand.

Crow Farm recommends the following flowers that are easy to grow on the Cape and make a nice bouquet: snapdragon, zinnia, cleomi, blue salvia, large marigolds, small sunflowers, black-eyed Susan, and orange cosmos.

Crow Farm is located at 192 Old King's Highway (Route 6A) in Sandwich. The retail stand is open May 1 through Christmas and offers annuals, fruits, flowers, and vegetables in season.

FOR MORE INFORMATION

call: 508-888-0690

TAYLOR-BRAY
FARM
YARMOUTH PORT

Tom Orton volunteers at Taylor-Bray Farm as a caretaker for miniature donkeys Nester and Sam.

This historic working farm allows visitors a peek into what life was like on the Cape centuries ago. The 22-acre farm was settled in 1639 by Richard Taylor and has a complicated history of ownership. The farm was in the Taylor family until George and William Bray bought the property in 1896. The Brays grew strawberries, blueberries, corn, and other produce, and harvested hay from the marshlands. After George Bray died in 1941, the farm changed hands several times.

The Town of Yarmouth voted to purchase it in 1987, "to maintain the farm for historic preservation and conservation purposes." It was placed on the National Register of Historical Places six years later. In 2001, the non-profit Taylor-Bray Farm Preservation Association and the Yarmouth Historical Commission joined hands to renovate and maintain the property for public enjoyment.

Today, sheep roam the pasture much as they did three centuries ago. A pair of Scottish Highland cattle, Scotty and Fiona, are also a presence in the fields. A flock of chickens live in the rebuilt chicken coops. As was the custom, an herb garden was planted outside the kitchen door for culinary and medicinal use. There is a community garden for townspeople to farm and tenant/managers live in the farmhouse.

OPPOSITE, CLOCKWISE:
A boardwalk leads to beautiful vistas where hay was once harvested.

Andy Rice of Hoggett Hill Farm, Brattleboro, Vermont, demonstrates how to sheer sheep, during the annual June Sheep Festival.

This sheep isn't sure what his future holds—a shearing or a Border collie.

Beth Deck demonstrates how to spin raw wool into yarn for Ruth Bray and her granddaughter Molly Maroney.

Taylor-Bray is a historical landmark that allows visitors a peek into life on a fam in 1639.

The grounds are used for Minutemen encampments, craft demonstrations, exhibits, and holiday celebrations. The grand fundraiser of the year, a Sheep Festival, is held in early June with sheepherding, wool spinning and weaving demonstrations, as well as hay wagon rides. Proceeds from parking donations help maintain the grounds and animal housing and helps feed the animals. The public is invited to visit and learn about the farm's Colonial past.

Tables are available for picnicking. There are short woodland walking trails and scenic views of Black Flats Marsh, where salt marsh hay was once harvested to feed the farm animals. The Sheep Festival is held in early June and a harvest festival is in the fall, with pumpkins, crafts, hayrides, and pilgrim interpreters. In early December, wagon rides, handmade holiday decorations, Christmas trees, baked goods, and mulled cider are offered.

TOP TO BOTTOM:
David Kennord of Wellscroft Farm in Chesham, New York, demonstrates the skill of his Border collies.

An elderly Border collie awaits a command in the shade.

CLOCKWISE:
The farm is open to the public without charge, but donations for the upkeep of the
buildings and care of the animals is appreciated.

Little boys can't resist a peek at the exotic llama and alpaca, who are members of the
Camelid family.

Hayrides are offered during the Sheep Festival and other special events.

53

When you are a miniature donkey, it's not easy to see over the stall.

Taylor-Bray Farm, 108 Bray Farm Road North (off Route 6A), Yarmouth Port, is open daily, year round from dawn to dusk. Donations are welcome and memberships in the Taylor-Bray Farm Preservation Association help preserve the farm.

FOR MORE INFORMATION CONTACT YARMOUTH HISTORICAL COMMISSION

call: 508-398-2231
membership information: 508-398-2231, extension 292
website: www.taylorbrayfarm.org

ELDREDGE FARM
BREWSTER

OK
BEETS
CARROTS

Eldredge Farm

Eldredge Farm, the former Red Gate Farm, is down a country road and in the woods on the Harwich-Brewster line.

Although Jeff Eldredge has a successful landscaping business, he yearned for a sustainable lifestyle. He wanted to grow vegetables and plants for the community. His wish came true when Red Gate Farm became available. The property produced fruit and vegetables for the neighborhood for generations. Now it is known as Eldredge Farm.

With an able manager, Steve Coleman, he took it a step further. Having 17 acres available and four greenhouses at his fingertips, the farm produces wholesale perennials, ornamental grasses, hydrangeas, shrubs, and trees, and retails to farmers markets all over the Cape. There is also a 50-member Community Supported Agricultural program.

The CSA runs 16 weeks, from June to October, if the weather cooperates. Each share receives a half-bushel basket of produce or about seven to ten items each week, which include herbs, leafy greens, alliums (onion family), fruiting vegetables, root vegetables, and, in spring and fall, one or two brassicas (members of the cabbage family). The memberships are sold on a first come, first serve basis. Crops are harvested on the morning of the pick-up day.

Eldredge has fine-tuned production so that it's as varied as a weather-dependent operation can be. June crops include salad mix, broccoli, radishes, Swiss chard, kale, spinach, Chinese greens, beets, carrots, cucumbers, tatsoi, pac choi, lettuce, peas, and spinach. In July, basil, beets, broccoli, cabbage, carrots, cauliflower, cucumbers, greens, onions, eggplant, bell peppers, hot peppers, yellow squash, zucchini, and radishes are available. August brings basil, beans, beets, broccoli, carrots, cucumbers, eggplant, onions, bell peppers, hot peppers, thyme, rosemary, yellow squash, and zucchini. In September, salad mix is available

Steve Coleman is the go-to guy and informal farm manager at the Eldredge Farm.

CLOCKWISE:

Four greenhouses make it possible to grow wholesale perennials, grasses, shrubs, and trees.

Customers who belong to the Community Supported Agriculture program pick up a half
bushel basket of produce each week, unless a disaster prevents it.

Indian Runner ducks provide a bit of entertainment on the farm for visitors.

This little kitten will grow up to be a great barn cat like its mother.

again, along with basil, beets, broccoli, corn, carrots, melons, onions potatoes, snow and snap peas, tomatoes, yellow squash, and zucchini. In October, it's salad mix, beets, broccoli, cauliflower, carrots, onions, potatoes, spinach, turnips, parsnips, yellow squash, and zucchini.

The landscape business provides lawn clippings and fall leaves for recycling and his livestock provide manure to create rich compost that is put into use on the farm. Speaking of animals, visitors will enjoy the friendly Buff geese, turkeys, free-range chickens (Orpintons, white cochins, sultans), along with humorous-looking Indian Runner ducks and pigs. If you've always pined for a pig, Jeff will grow one for you.

LEFT TO RIGHT:

Tomatoes are the most popular crop on the Cape.

Perennials are sold to nurseries, garden clubs, and at farmers markets.

Carol Carey, who works in the greenhouse, holds a Thanksgiving turkey that will belong to some lucky person.

58

This shed is the pickup point for Community Supported Agriculture customers.

Steve Coleman works the fields as well as the markets.

Eldredge Farm, 24 Eldredge Farm Cartway, Brewster, is located off Depot Street on Squantum Path. Drive 100 yards and bear left at the fork to Cranberry Trail. Continue on to the entrance to the greenhouses and fields. Open until 7 p.m., with assorted plants, produce, and eggs. Products are sold wholesale to local nurseries and garden clubs and retailed at farmers markets throughout the Cape.

FOR MORE INFORMATION

website: www.eldredgefarm.com

The Coonamessett Farm encompasses twenty acres, but only half is farmed.

COONAMESSETT FARM
EAST FALMOUTH

Ron and Roxanna Smolowitz don't have romantic notions about farming. Both are scientists and work to support the farm, rather than the other way around. They have owned and operated Coonamessett Farm since 1989 and have been committed to its care. They do it because they believe in the concept of growing food locally. They are holdouts from a world that wants everything yesterday.

As a veterinarian, Roxanna cares for the many animals on the farm, but she is also a visiting assistant professor in Marine Biology at Roger Williams University in Rhode Island, as well as the director of the Aquatic Diagnostic Laboratory. Ron is an engineer who retired from the National Oceanic and Atmospheric Administration (NOAA) but still consults and conducts research.

The property was once part of Coonamessett Ranch, an enormous dairy farm that went broke at the turn of the twentieth century. The land was divided in 1935. Coonamessett Farm encompasses twemty acres, but only half is farmed. The work is labor intensive, weather dependent, and risky, but few people experience the level of satisfaction that this couple does from farming the land. In recent years, they have added to their greenness with a wind turbine and solar panels.

The Smolowitzes are community-minded and offer a program called Little Sprouts, where children grow vegetables, herbs, and flowers in their own plot and learn about organic farming, insects, worms, and composting. A summer harvest celebration party is held at the end. Recently, they built a Montessori-based middle school

Field manager Stan Ingram tends to the hydroponic lettuce in Greenhouse Five.

Farmer Ron writes a monthly e-mail newsletter for members with free advice. But you must take the Membership Oath:

"I pledge to obey all the farm rules that have been posted in inconspicuous locations throughout the farm as well as those rules found in M.G.L. c. 1786 C.M.R. 151(9)(a) or any rules found laying around unused. (Consult an attorney before signing.)"

TOP TO BOTTOM:

In the winter, lettuce is grown hydroponically in the greenhouses to defray the cost of operating the farm. Later, seedlings are planted in the fields or sold in six-packs to the public.

In recent years, the Smolowitzs added to their greenness with a wind turbine and solar panels.

classroom called the Coonamessett Farm School, which promotes stewardship of the environment and appreciation of the role of agriculture in society.

Three different Community Supported Agriculture programs are available: produce, meat, and cheese. The cheese comes from farms in the eastern part of the state, and the meat from the Farm Institute on Martha's Vineyard and other nearby farms. They also have a farm membership program which includes unlimited visits with the farm animals and discounts on farm events and rental fees on kayaks and canoes, which can be used on a nearby pond.

There are six greenhouses in use (including one used to grow lettuce hydroponically), raised beds, cold frames, a memorial herb garden, and a spinning windmill. Plus plenty of animals: alpaca, miniature Mediterranean donkeys, Nigerian dwarf goats, a tortoise, Baby Doll sheep, Shetland sheep, assorted ducks, chickens, turkeys, usual birds, and a regular dog.

The general store carries jams made by Roxanne, authentic alpaca yarn, unusual gifts, fresh produce, and old fashioned candy. There is also a cafe with indoor and outdoor seating that overlooks the farm and an ice cream parlor.

Visitors can enjoy breakfast or lunch while taking in a view of the farm.

The Coonamessett Farm general store carries jams made by Roxanne, authentic alpaca yarn, unusual gifts, and fresh produce.

INSETS:
A barnyard cat takes a rest from her duties.

A dust bath is better than no bath for this miniature Mediterranean donkey.

What's a general store without old fashioned candy?

Coonamessett Farm, 277 Hatchville Rd., East Falmouth, is open seven days from 9 a.m. to 5 p.m., from late April through Christmas Eve. School tours are available, as well as internships, 4H Program, special events, parties, plants, produce, lunch, and gifts.

FOR MORE INFORMATION ON SEASONAL ACTIVITIES, SUCH AS THEMED EVENTS LIKE A CAJUN BUFFET AND DANCE OR SUSHI WORKSHOP.

call: 508-563-2560
website: www.coonamessettfarm.com

CAPE COD
LAVENDER
FARM
HARWICH

OPEN HONOR SYSTEM

Planting, weeding, and harvesting is all done by hand.

Many people don't realize there is a 20-acre lavender farm right in Harwich Center. The sign on Route 124 is small and unassuming, but those who follow the unpaved winding road to Cape Cod Lavender Farm will find themselves returning. Where else can you walk through fields of lavender on a summer day and have the fragrance linger for hours? Where else can you find a shimmering oasis of purple flowers surrounded by conservation lands and woodland trails? Nowhere, because it's the only lavender farm on the Cape.

Owner Cynthia Sutphin discovered that the climate of the Cape is similar to that of Italy, southern France, and Greece, where the plants grow wild. She became a farmer in 1995 because she wanted a business at home so she could raise her four children. They grew up working in the fields with her.

Sutphin started out going to local farm markets with bunches of fresh lavender. As her business grew and flourished, her farm became a destination that has been featured on televisions shows like "Rebecca's Garden" and "Chronicle," and in magazines, newspapers and websites.

It is a romantic, beautiful place, where the air is filled with the songs of birds. But it's not a resort. Planting, weeding, and the harvesting of the delicate flowers is all done by hand. And then there is year-round maintenance. Still it's an enviable lifestyle for those who love nature and solitude. Even the bees are happy as they wallow in the abundant nectar.

Cynthia Sutphin started Cape Cod Lavender Farm so she could be home with her children while they were growing up.

Sutphin has suffered through years when winters were unkind to her plant stock. But she says she would do it over again in a heartbeat. She has no intentions of getting any bigger. Instead of expanding, Sutphin has refined what she has.

One year she created a woodland fairy castle garden for children with local sculptor Eddie Foisy. Another, she planted a peony garden as a special place for the family to relax. Thoughts of a tree house dance in her head. But mostly she just wants to maintain what she has with her husband, boat captain Matt Sutphin. For her, living on a lavender farm is normal, but she is well aware of how rare the lifestyle it is. Both of her daughters had their weddings there and she is entertaining ideas of opening the property for others. Her biggest wish is that one day it will become a large family compound.

LEFT TO RIGHT:

The lavender is at its peak in late June and early July.

Workers carry their morning harvest to the shed to be bundled for sale.

Cut lavender, lavender products, and plants are sold in the shed.

Each delicate lavender blossom is cut by hand.

Sumo, the black Lab, is part of the Sutphin family.

Sutphin commissioned local sculptor Eddie Foisy to create a fairy garden in a shaded nook
so that children have a special place at the farm.

Both of Sutphin's daughters chose to have their weddings at the farm.

The shop is open year-round on the honor system.

Workers return year after year. The farm is about family and workers who become part
of it, like Liz Holtman who returned for five seasons, Cynthia Sutphin the owner, niece
Malaney Cassano, and daughter and store manager Anna Ramsay, who has worked at the
farm every summer since she was 10.

Tips from the top

Lavender is delightfully fragrant, whether fresh cut or dried.

Water carefully once a day for two weeks when lavender is first planted. After it is established, only water if the plant wilts. Lavender loves sun, sandy sweet soil, dryness, and air around it. Lavender grows in partial afternoon shade but it's at its best in full sun. Fertilize lightly with cow manure mixed with cottonseed meal. Spread around the plant and scratch it in. Top dress with clean white sand (not beach sand), can prolong the life of the plant by cooling the roots, improving drainage, and reflecting light onto the plant. The biggest problem is fungus from over-watering.

Lavender can be added to flower arrangements to make them more fragrant. Dried, it can be mixed into potpourri and sachets or added to foods in place of rosemary. Some claim it's an aphrodisiac, others, a moth repellent.

Cape Cod Lavender Farm is off Route 124 at 41 Weston Wood Rd., Harwich Center. Follow Island Pond Trail to the lavender fields and shed where plants, soaps, lotions, sachets, candles, and other products are available. Fresh cut lavender is for sale in June and July during harvest time.

FOR MORE INFORMATION OR TO ARRANGE A GROUP TOUR

call: 508-432-8397
website: www.capecodlavenderfarm.com

The FEED BARN & NURSERY

BLUE SEAL

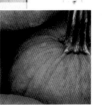

Coca-Cola

Phil Roy manages the Marston Mills farm, where crops are grown for the Cape Abilities Farm stand in Dennis.

Cape Abilities Farm started out as a wonderful concept in 2006 and became a successful enterprise. The farm, a division of a non-profit that provides jobs, homes, transportation, support, and other services for people with disabilities on the Cape, began with five paid employees. It has grown exponentially since then.

Within two years, a farm in Dennis was donated to the organization and then another in Marstons Mills. With eight acres and six greenhouses, the crop list has expanded and so have the job opportunities, from planting seeds to packaging the award-winning tomatoes, squash, greens, herbs, and other produce. The farm is different from other farms in that it has accessible asphalt paths outdoors and extra wide spaces between greenhouse rows to accommodate wheelchairs.

Cape Abilities Farm now has 80 employees with disabilities, who work in all aspects of farm management. Employees work year round, planting seeds and nurturing them under grow lights until they are ready to move to the greenhouse. The greenhouses are high tech with hydroponic watering systems, which allows them to stretch the growing season to eight months.

Seeds aren't the only thing that is nurtured in the organization. Each employee has a job coach who stays with him or her until a skill is mastered. A worker's satisfaction

TOP TO BOTTOM:
Workers take pride in their work, which provides the community with food.

The wide open space of the fields provides maximum sunshine for vegetables.

OPPOSITE:
With a lake nearby, the farm is a pleasant place to be on a summer's day.

Tools are neatly lined up inside the workshop.

goes beyond a job well done. By providing food, plants, and flowers to the community, they feel like an important part of it.

Produce grown on the historic working farm in Marstons Mills is brought to the farm stand on Route 6A in Dennis and sold. The Salad Club (a CSA program) is offered to those who pay a fee at the beginning of the summer and then pick up weekly grab bags of fresh produce. Fine restaurants, like The Naked Oyster in Hyannis, Red Pheasant in Dennis, and Brewster Fish House, also feature Cape Abilities produce on their menus.

Tomatoes are the biggest crop and 30,000 pounds are produced hydroponically in the greenhouses annually. They have won the Boston Tomato Festival contest year after year and it has become their signature crop. Other products, such as soaps, jams, jellies, salad dressings, and honey that are locally made or grown, are also available at the stand.

With the donation of this old farm in Marstons Mills, the crop list has been expanded as have the job opportunities.

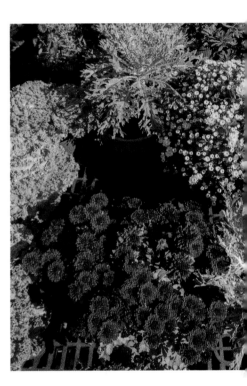

LEFT TO RIGHT:

Workers have the joy of seeing the seeds germinate, then blossom, like this squash flower.

Squash is grown on the farm, then brought to the farm stand for sale.

Mums and ornamental kale sold at the farm stand will help beautify customers' gardens.

In pumpkin season you can take your pick at the Cape Abilities Farm Stand in Dennis.

Cape Abilities Farm Stand, 458 Main Street, Route 6, Dennis, is open from 10 a.m. to 5 p.m. daily from May to December. Hydroponic tomatoes, butter-crunch lettuce, romaine, oak leaf, arugula, salad bowl, gold and green zucchini, summer squash, eggplant, bell and chili peppers, and cucumbers are usually available. Look for field vegetables, flowering annuals, and perennials in season, as well as pumpkins in the fall and Christmas trees and wreaths in winter.

FOR MORE INFORMATION

call: 508-385-2538
email: farm@capeabilities.org
website: www.capeabilities.org

E & T
FARMS
WEST BARNSTABLE

Sunflowers are a popular choice at the farmers markets where the Osmons sell.

Ed and Betty Osmun operate the only combination hydroponic/aquaponic farm on Cape Cod. As if farming weren't difficult enough, they raise fish as well. And they have a passion for bees. They tend 200 hives from one end of the Cape to the other, and also rent hives to farmers for pollination. E&T Farms, named for Ed and son Ted, specializes in hydroponic lettuces, edible flowers, spicy Asian greens, and herbs, as well as tilapia, koi, and trout.

It came about because they took a trip to Disney World and saw how hydroponics could be combined with aquaculture to create a sustainable system. They didn't just admire it, they dreamed of their own. It was an efficient use of land, which is at a premium on the Cape. And they could grow year round.

In 2002, they left their day jobs. Ed built a 9,500 square foot indoor growing space and created nine indoor fish ponds that connect to the greenhouse. It's a highly complex system that only Ed understands, involving biological filters, air systems, raceways, tanks, and hydroponic trays. Thankfully he is blessed with strong mechanical skills.

The system works like this. The nutrient-rich water the fish are grown in is circulated through raceways into the greenhouse trays, where it the feeds the plants. The plants in turn filter the water by removing the nutrients. The water is then filtered again by spinning wheels before it is returned to the fish tanks. Simply put: the fish feed the plants and the plants clean the water for the fish. It's a perfect circle of sustainability.

Ed and Betty Osmon are high school sweethearts who worked for 30 years in other fields before pursuing their dream of hydroponic farming.

The koi are sold to garden centers and to the public, the bass are used to stock ponds, and the tilapia go to the Asian food market in Boston. They leave the premises alive and whole in tanks or packed in ice. Produce is sold at health food stores, farmers markets, and at their farm stand in West Barnstable. It's also featured on the menu of local gourmet restaurants.

They have recently added chickens to the 5.5-acre farm and planted an acre with blueberries, WWwhich they plan to maintain in the winter when they are not busy with their bees.

CLOCKWISE:

Ed Osmon cleans his garlic crop during a lull at the market.

Yellow is a favorite color for vegetables as well as flowers.

A bluebird of happiness visits the farm.

Betty Osmon makes beeswax candles and tends the stand
alongside her husband.

E&T Farms, 85 Lombard Ave., West Barnstable, is open year round, weekdays 10 a.m. to 4 p.m., and Saturday after 2 p.m., with hydroponic vegetables, herbs, blueberries, garlic, basil, honey, honey pollen, candles, beeswax products, and koi.

FOR AVAILABLE TOURS & MORE INFORMATION

call: 508-362-8370
website: www.eandtfarmsinc.com

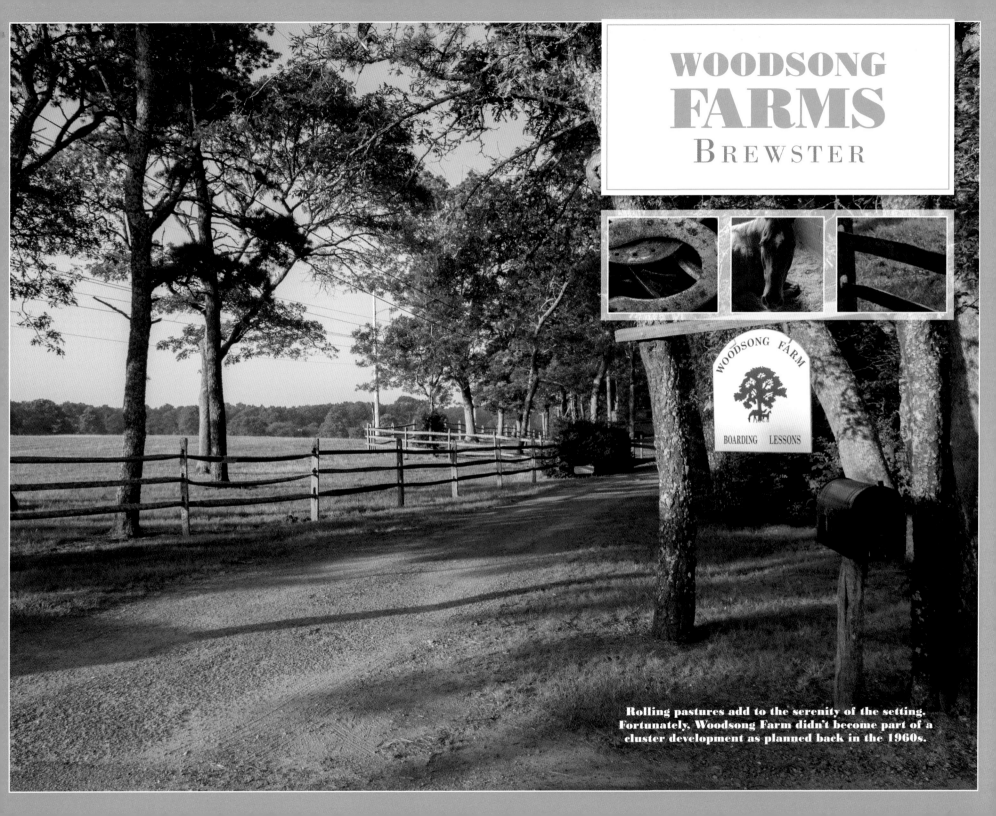

WOODSONG FARMS
BREWSTER

WOODSONG FARM

BOARDING LESSONS

Rolling pastures add to the serenity of the setting. Fortunately, Woodsong Farm didn't become part of a cluster development as planned back in the 1960s.

Marvelous rural areas still exist on Cape Cod, but they are often hidden from the main roads. This peaceful seven-acre horse farm in Brewster is an example. Once part of the 365-acre Pine Ridge Dairy Farm, developers bought it in the 1960s, along with an adjacent turkey farm and nine-hole golf course, and planned to turn it into the largest cluster development ever seen on this side of the bridge. But a young gal named Bette Avery happened along.

The Cape Cod native was home from college and working as riding instructor at the former Sea Pine School in Brewster. One day while riding through the area, an idea came to her. The 21-year old approached the owners of Cape Land Inc. and asked if she could lease some of the land as a horse farm.

In 1967, the persuasive young woman signed a five-year lease for $1 a year and secured a $10,000 loan to buy horses and equipment and replace the old cow barns and pig house. She's been teaching dressage and hunter/jumping, boarding, coaching, training, and enriching the lives of her students ever since.

Avery became enamored with horses after receiving one for Christmas when she was 9. It was all she ever wanted. Later her parents stood by her side when she took on the responsibility of the horse farm, helping her gut and rebuild the barn and clear a ring for riding. Finally, in 1980, she bought the property.

Bette feeds a flower to a favorite horse.

Each horse is watered and fed three times a day and turned out for fresh air.

Improvements have been added over time and the farm is now a year-round facility, thanks to more people with means living on the Cape. Some of her students became outstanding riders who went on to take national awards. It's a surprise to her that her five-year commitment has become a lifetime. But she can't imagine living any other way.

Every day, after breakfast, she feeds and waters the ten horses in the stables and turns a group of them outside. Then she cleans the stalls. After that it's time for lunch. For the horses that is. She brings in the ones that are out and brings out the ones that are in. Then she has lunch herself before her classes begin at 3 p.m. Night chores await her when she finishes. She has to stable the horses, feed them, clean the stalls, and catch up on paperwork. She finally gets to bed after midnight.

There are no holidays, vacations, or sick days. But she loves it and gets satisfaction out of her work, even if her freedom is limited. Her children are involved in the business as well. Her son helps out as a handyman and her daughter teaches and acts as a coach and trainer at competitions. She has also started a middle and high school riding team similar to intercollegiate competitions; her young charges have made it to regional finals. It's a different kind of competition because the students ride a horse they have never been on before.

Avery is also involved in the City to Saddle program, a non-profit charitable organization for inner city children. As a host facility, she provides spaces in her summer riding camp for disadvantaged children who normally wouldn't have the opportunity to ride.

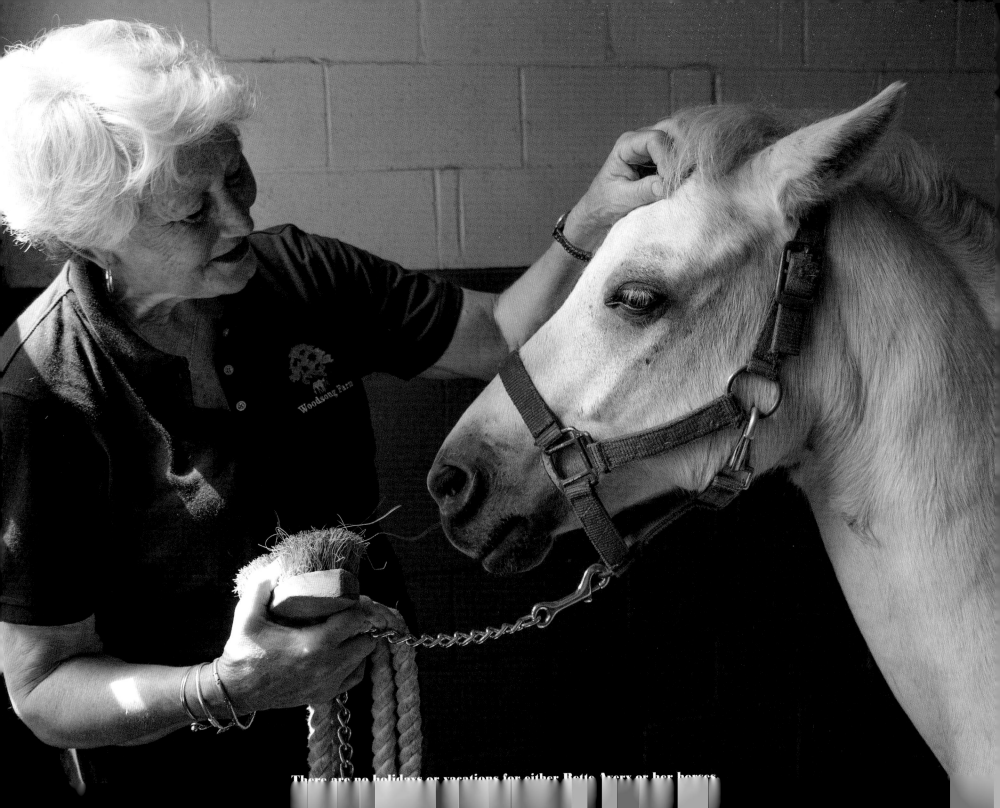

There are no holidays or vacations for either Bette Avery or her horses.

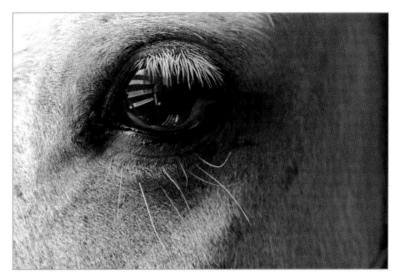

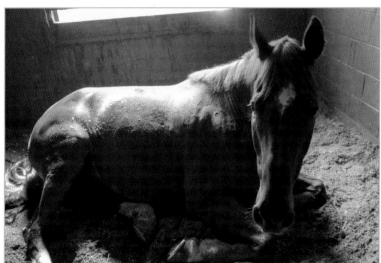

CLOCKWISE:

The stable is reflected in the eye of the horse.

Never throw a lucky horseshoe away.

Mist rises from a meadow in the early morning.

Even horses get tired of standing on their feet all day.

Children enjoy the horse shows that are scheduled each month at the farm from May through October. Here, Lea Kilpatrick readies her horse for a trial.

Woodsong Farm, 121 Lund Farm Way, Brewster, open year round. Besides a 24-stall horse stable for boarding, it offers horses for sale, breeding, riding, equestrian lessons, horse shows, training, and a fully stocked tack shop. A horsemanship day camp program, Pony Kids, is offered in the summer.

FOR MORE INFORMATION

call Bette Avery: 508-896-5800

CHECKERBERRY
FARMS
ORLEANS

The Norgeots collect antique farm tools for fun.

Gretel Norgeot has grown food her whole life. As a child, she helped her grandmother and her mother in their vegetable gardens. When she married and had her own children, she started her own seven-acre farm. Over the years, her experience has grown and her knowledge has increased. Norgeot became a farm market manager, master gardener, farm activist, and leader in her community.

As a young mother in 1995, she was an original vendor at the Orleans Farmers Market. Every week she arrived with honey, eggs, produce, hand-spun yarn, and three children. It was something they all could share.

Norgeot learned the importance of good quality and consistency first hand. When the market founder, Tina Van de Water, moved away, Norgeot stepped up to the plate. She became responsible for signage, advertising, and bookkeeping while being a vendor herself. She believed strongly in the market and felt it was a benefit to the town. It gave her the courage to deal with the board of health and selectmen. She also went on to incorporate the market and find a permanent location for it.

When Norgeot joined the Mid-Cape Market as a vendor, she ended up managing that as well and also took on a third market in Provincetown for a year. She also initiated a vegetable garden at Stony Brook Preschool for two years until the space became a baseball diamond.

Joan Hitchcock (middle) taught her daughter Gretel (right) to farm and now she has taught her daughter Marie (left), so three generations of women are farming together.

LEFT TO RIGHT:

This red rooster wakes the family at dawn and also keeps the hens in line.

Marie Norgeot farms her own plots at the family farm and now operates a small Community
Supported Agriculture program for a few neighborhood families.

Garlic scapes have become a popular gourmet treat at the farm markets.

Norgeot believes strongly in the merits of farming and approached Nauset Regional High School about developing a farm program. She went through an agriculture program at a technical high school herself so she knew its lifelong value. With her enthusiasm and energy she got lumber donated, as well as soil. Then, with her husband and friends, she built raised vegetable beds so students could grow salad greens for the school cafeteria. Science teachers oversee the program under her direction and it's been a success.

On her own farm, she grows what she likes to eat. That means lettuce, carrots, cucumbers, beans, peas, beets, and squash, along with perennial mainstays like rhubarb, strawberries, and asparagus. She also grows a tremendous amount of garlic each season—enough to fill her shed. Her daughter Marie is actively involved in the farm; she provides a CSA for a dozen local families and does the vending at markets.

The Orleans market has almost tripled in size under her guidance. Even with a board of directors and committees to help with the work, she is still the go-to person. Norgeot's role has become bigger than just being a farmer. She has become a farm activist.

She and her husband were on the policy committee for the National Farmers' Union; she attends market manager conferences on the state level and is on the board of directors of Buy Fresh, Buy Local and the Orleans Agriculture Council. She hasn't lost her personal touch, however. One of her favorite activities is to give educational tours of Checkerberry Farm to school groups.

CLOCKWISE:
Norgeot grows enough garlic to fill her shed at drying time.

A handful of fresh garlic goes a long way.

Each time Norgeot bought more chickens she chose a different type, and her eggs reflect the difference.

Norgeot shares her knowledge about raising chickens with Lori Caron of Salty Lou's Lobsters, a fellow vender.

It's home sweet home as Buddy waits for his mistress to
return from the fields.

Checkerberry Farm, 46 Tar Kiln Road, Orleans, has
a roadside stand. CSA available. Products include seasonal
produce, honey, eggs, and yarn. They can be found at the
Wellfleet Farmers' Market on Wednesdays.

FOR MORE INFORMATION

call: 508-237-9492
email: greteln@capecod.net
website: www.checkerberryfarm.com

TRURO
VINYARDS
TRURO

The entry to the winery features a wooden barrel garden.

TRURO
VINEYARDS
OF CAPE COD

Most people think of sandy dunes, salty air, and pristine beaches when they think of Cape Cod. They don't think of grapes. That's why Truro Vineyards is so unexpected. Located a mile from the ocean and bay on the outer reaches of the Cape, the six-acre property takes advantage of its resemblance to Northern France. In climate that is.

The location is blessed with warm temperate breezes, even moisture, sandy soil, and sunny days. It's the perfect combination for growing grapes loaded with flavor and the lush varietal character that goes into the vineyard's merlot, chardonnay, and cabernet franc. While some grapes are imported from Massachusetts, New York, and California for blends, only those grown on the property go into their estate wines. All the wines are made and aged in their barrel room.

Park in the white shell driveway off Route 6A and walk around the peaceful grounds. The Federalist farmhouse, now used as a wine shop and gift store, dates back to 1830 and was once painted by the artist Edward Hopper. An ancient Chinese mulberry tree, brought over by a sea captain, provides restful shade. In the 1990s, the land, which is close to 44,600 acres of Cape Cod National Seashore, was transformed into a vineyard.

When the Roberts family bought it in 2007 and expanded the business, the vineyard finally reached local prominence. Dave was in the fine wine and spirits industry for 40 years before he retired to pursue his dream. The growing interest in locally made products and the popularity of wine helped draw attention to the vineyard, but it is the energy of this hard-working family that has made it a success.

Winemaker Matias Vogel shows Rebecca Gawley how to gauge the ripeness of the grapes.

David and Kathy Roberts, their children, Kristen and Dave, Jr., and his wife Amy, actively run the vineyard. Although they modernized the facility, and tripled the size of the area, they want it to remain small. They strive to improve their techniques in both the fields and barrel room. Their focus is on quality, not quantity. Matias Vogel, a winemaker from Budapest, works with Dave, Jr., who has become a master as well.

TOP LEFT:
The vines cast shadows to infinity.

LEFT TO RIGHT:
This giant mulberry tree was brought to the Cape by a ship captain and is a focal point of the vineyard.

The public is free to explore the grounds and enjoy a picnic.

Grapes are covered with netting to keep wildlife from temptation.

BOTTOM RIGHT:
Warm days, sandy soil, and even moisture infuse these grapes with flavor.

97

The handcrafted wines include Sauvignon Blanc, Chardonnay Vignoles, Cape Blush, Cabernet Franc, Maritime Red, Triumph Cranberry, and Diamond White. In all, they produce 13 different wines including blends called Right White and Right Red, which help benefit the Provincetown Center for Coastal Studies. In 2011, Diamond White was named the best wine in Massachusetts at the Eastern States Exposition.

The Robertses are proud of what they have accomplished and invite visitors to see how they do it. The yesteryear quality is still evident, even in the state-of-the-art winemaking facility, which is lined with oak barrels that seem to have been there forever. Visitors can stroll the vineyards, use the picnic grounds, taste varietals and blends, and shop for local crafts, wine accessories, and unusual items in the quaint gift shop. Tastings are held outside in an open air tasting pavilion. Next on the agenda is a new distillery, where they will make rum.

Tours are free and are held at 1 and 3 p.m. every day from Memorial Day to Columbus Day weekend. Each September, the vineyard opens its doors and vats to the public for their annual Grape Stomp and Jazz Fest. Volunteers from the community pick grapes and help coordinate the event. And yes, you can actually stomp grapes as well as enjoy food, wine, and music.

The climate of the Cape is similar to that of Northern France which makes it perfect for great grape growing

OPPOSITE. CLOCKWISE:
Wine tastings are held daily.

Manager Rebecca Gawley toasts a customer. Tastings are held on the outside patio.

Truro Vineyards produces 13 different wines including blends called Right White and Right Red. A portion of the sales benefit the Provincetown Center for Coastal Studies.

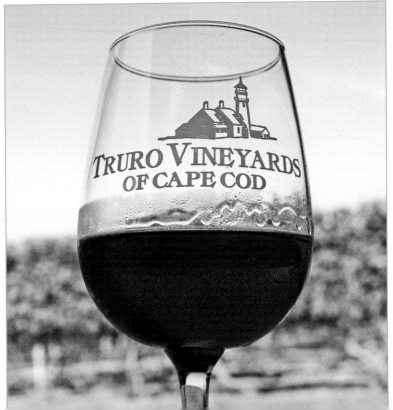

The ancient Chinese mulberry tree provides shade.

Truro Vineyards of Cape Cod, 11 Shore Rd., Truro, is open daily 11-5 p.m. and Sundays 12-5 p.m. through October, and Friday to Sunday in November. Tastings are held every half hour.

FOR MORE INFORMATION OR DATES OF SPECIAL EVENTS

call: 508-487-6200
website: www.trurovineyardsofcapecod.com

TOMATOES $3.50/lb.

Cherry tomatoes
$4.

Organic herb plants at Bay End Farm.

David Ingersoll decided he wanted to raise his family in a healthy environment, so he took advantage of an unusual situation. This six-acre farm was originally part of Grazing Fields, a 900-acre dairy farm purchased in 1906 by his great grandmother. She ran for president of the Farm and Labor Party and was more of a manager than a farmer. But she loved the lifestyle and put an agricultural restriction on the land so it would be preserved when she died. Her daughter took up the reins, growing her own food and canning for winter. She managed to live off the land, but when she passed on it went unused for a generation.

Until her grandson David, or Kofi as he is called, came along in 1998 and decided to take up the challenge. Born in West Africa and raised in Cambridge, he and his wife Erin started out farming two acres organically. Now they are up to six. The rest of the land is held in trust and can't be sold without all the family members approving. So far, he is the only one interested in farming.

Their children Maya and Zeke are now growing up on the farm. They play in the wide open spaces, often sampling the fresh peas or cherry tomatoes their parents are harvesting. The Ingersolls grow as many different crops as they can, including a variety of lettuces, arugula, chard, spinach, snap peas, cucumbers, green beans, potatoes, red and yellow onions, red carrots, napa cabbage, bell peppers, eggplant, kohlrabi, edamame, leeks, kale, baby bok choi, summer squash, and fresh herbs. Heirloom tomatoes are their biggest crop and they grow flowers for weddings.

In addition to a farm stand, they offer a seasonal CSA to about a 100 families. It runs for 20 weeks from mid-June through the end of October. Customers receive a weekly

Kofi Ingersoll, daughter Maya, and wife Erin come in from picking in the fields.

bag of fresh, seasonal, organic vegetables, herbs, and a bouquet of flowers. Each bag feeds a family of four for a week. They also sell their produce to Cambridge restaurants.

Unfortunately, farming only provides the young family with half an income. To support themselves, they also manage another property nearby and rent it out for weddings.

Kofi says the state has seen a 25 percent increase in the number of new farms during the last decade and feels that he is doing something meaningful with his life. He remembers picking corn on the day the Twin Towers came down in New York. He feels people still have to eat, and growing good, healthy food for them makes sense. He also thinks it's a great way to raise his children and get them involved in the natural world.

LEFT TO RIGHT:
Bags of produce await pickup by Community Supported Agriculture members.

Community Supported Agriculture customers receive a bag of fresh picked, organic vegetables and herbs each week that will feed a family of four.

Customers are invited to take flowers on an honor system.

103

The farm is Maya Ingersoll's playground.

Erin Ingersoll teaches her daughter by letting her taste the produce.

Bay End Farm at 200 Bournedale Rd., Buzzards Bay, is one mile from the Bourne Rotary. Open mid-June to mid-October on Thursdays from 2 to 6 p.m., and Saturdays from 10 a.m. to 2 p.m. CSA is offered and there are several group pick up locations with deliveries to Plymouth and Cambridge. Work plans are available.

FOR MORE INFORMATION

call: 617-212-8315
website: www.bayendfarm.com

KANE
FARM
TRURO

Kane always knew he would return home to farm his land.

Tom Kane grew up on his grandmother's farm, where she grew the food they ate and preserved enough for winter. Times were tough in the 1940s, but his grandmother was tougher. She taught him how to skin a rabbit and cook it for dinner. As a boy, Tom hunted, clammed and fished, and, in time, passed his knowledge along to his own sons, including how to dress a rabbit.

The Kane family goes back five generations in this small town. His uncle and namesake was a well-known columnist for the local paper and a farmer, which his own dad wasn't. Uncle Tom once owned part of the property but traded it for a down payment on a house.

When Tom grew up, there was no work available on the Cape. He left to join the Navy and later to become an electrician. But he always knew he would return and farm the land. He finally did it in 1973, and has been happy as a proverbial clam ever since.

Tom and his sister, who lives in the upper section, divided the land between them. Today, Tom's son, also an electrician, lives in the compound as well.

It took several decades to get the property the way this Kane wants it, and he wants it picturesque and neat as a pin. When a former kettle pond started to fill in and become a breeding ground for mosquitoes, he dug it out and created a man-made fish pond.

Although he grows vegetables in quantity, he doesn't sell them. He gives them away to friends and customers. Or trades them for pies and cakes. He claims he's the only electrician around who gives his customers cucumbers and squash with a service call.

Farmer Kane, an electrician turned farmer, is up at 5 a.m. to feed the sheep and pigs, and then himself, in that order.

COUNTER CLOCKWISE:

Three generations of Kanes live at the farm.

A turkey struts his stuff in the barnyard.

Baa, baa black sheep, have you any wool?

Kane loves his animals, but eats them too.

Tom farms as a hobby. But he also has a thriving business growing Alpine succulents with a partner who sells them to nurseries from the Cape to Boston. He calls them high class weeds because they love drought and poor soil, of which he has plenty.

He orders plants from all over the world, and Oregon in particular. To keep them happy, he serenades them with music. Only after studying them for a year will he propagate them for sale.

He also raises pigs and lambs for people he knows as well as himself. He says he loves his animals, but he eats them too. With the help of a trekker from Nepal who visits every year, the farm looks more like a park with a petting zoo. One year, they built an unusual farm stand that resembles a shelter in the Himalayas. Tom puts out buckets of sunflowers that he grows for sale and an occasional carton of eggs or jar of honey, but he says he can't make money farming.

He's up at 5 a.m. to feed the ducks, sheep, and pigs, and then feeds himself—the way his grandmother taught him. After that it's a mix of farm chores and electrical work. It adds up to a 12-hour day, but the four-acre farm is his pride and joy. His reward is eating nothing but the best.

Succulents are easy to grow, if you have the time and patience.

INSET:
When Kane learned how to propagate plants, it turned into a business.

When a kettle pond filled in and become a breeding ground for mosquitoes, Kane dug it out and created a man-made fish pond.

TOP TO BOTTOM:
Kane sells plants to nurseries from the Cape to Boston.

It took several decades to get the property looking the way he wanted it – neat as a pin and pretty as a picture.

Tips from the top

Original plants are used for propagation and never sold.

When propagating don't strip the mother plant bare. If there are seven babies, leave three.

Take cuttings from new growth—pieces that are young, tender and healthy.

Be there and be persistent.

If you see flies around you have aphids, because flies eat aphids.

To grow plants, you have to love them and have a green thumb.

Kane Farm, 2 Hatch Rd., Truro, sells eggs, honey, and sunflowers when available. Occasional seasonal vegetables from guest vendors from other local farms are available.

FOR MORE INFORMATION

call: 508-349-6283

PLEASANT LAKE
FARM
HARWICH

Chickens who eat marigolds have brighter colored yolks.

Robinson "Skipper" Lee grows certified organic vegetables and meat on his 12-acre farm in Harwich. What is unique about the farm is that Lee, a tug boat captain in Boston, has harnessed the sun and the wind with solar panels and a wind turbine. His goal is to take nothing from the environment and as a result, he lives totally off the grid.

No chemical or synthetic fertilizers are used on the certified organic farm, as fertilizer is provided the old fashioned way by the picturesque Scottish Highland cattle. Lee bought the 15 acres in 1997 when it was slated to be a development. He sold three of them and placed the remaining acres under an agricultural restriction. While erecting a cement block house with the help of Cape Cod Regional Technical High School students, he lived in a barn he built on the property. His utility bills are nil due to a 1,000 watt wind turbine that is 100 feet tall with a wing span of 10 feet. His central vacuum runs off the solar and wind power and he uses propane gas for the refrigerator. His home is heated by a wood and coal stove. But there is no television.

Lee started out farming with a specialty crop of 16 varieties of garlic. It was a first and so successful that other farmers soon copied him. Since then he has expanded his operation to include potatoes, tomatoes, snap peas, chard, greens, beans, summer and winter squash, heirloom lettuces, field greens, mustards, and herbs. Edible flowers, such as dianthus, bachelor's buttons, nasturtium, borage, and chrysanthemum, add color to the farm stand as well.

A fruit orchard is planned where certified organic cherries will grow. He also offers a CSA with the help of his farm supervisor, Veronica Worthington. School groups are invited to visit (by appointment) to see what an organic

Skipper Lee and Rebecca Scanlan have created an old-fashioned farm lifestyle at Pleasant Lake Farm in Harwich.

CLOCKWISE:

Pleasant Lake Farm is totally off the grid. No television before bed for this farmer.

A tugboat captain and mechanic needs a good-sized barn to operate.

The farm is hidden in the woods, so it takes a little patience to find it.

The farm is named for a nearby lake.

113

farm and alternative energy is all about. His strong suit is that he can get any machinery to run on the farm, including vintage tractors.

One of the biggest problems with farming for Lee is finding help in season, as there are higher paying jobs elsewhere. So he looks for people who are in a position to do it, who want more out of farming than a paycheck. He would also love to see more people farming on the Cape again, as he is.

LEFT TO RIGHT:

Danielle Pattavina is learning a lot about organic farming while working at the farm.

Pattavina protects herself from the harmful rays of the sun.

Lettuce is mulched with nutrient rich seaweed.

TOP TO BOTTOM:
The first pair of Scottish Highland cattle came from Taylor-Bray Farm in Yarmouth Port.

Scottish Highland cattle provide fertilizer for the farm as well as meat for the freezer.

LEFT:
Only natural light and natural food at this farm stand.

115

Tips from the top

Garlic keeps up to six months and shouldn't be refrigerated.

Garlic likes a pH of 6.5. Lime the soil if the pH is low,

Garlic should be planted in late October or early November.

Buy organic seeds if possible. Supermarket garlic is mostly grown in China, then cured and sprayed to keep it from sprouting.

Mulch the planting with leaves or straw over the winter. In spring uncover it partially so the sun can warm the soil. When it sprouts, remove all the mulch.

Garlic needs water early in the season, so irrigate during a dry spring.

Add nitrogen to encourage the green tops to grow. Called garlic scapes, they are a delicious crop in themselves and can be harvested in spring. A dry spring will bring an early harvest and smaller bulbs. Harvest the bulbs in early July. (Varieties vary from early to mid and late season.)

Garlic keeps up to six months depending on the variety and whether the conditions are dry. It shouldn't be refrigerated.

The Pleasant Lake Farm Stand, 2 Birch Rd., Harwich, (left off Azalea in the Headwaters region) opens in June, Wednesday to Sunday from 10 a.m. to 5 p.m., with certified organic vegetables, edible flowers and herbs in season. The garlic crop will be available in late July. School groups are welcome by appointment.

FOR MORE INFORMATION

call: 508-878-9146
website: www.pleasantlakefarm.com

HOLLY
FARM
BREWSTER

This holly, *ilex pedunculosa*, looks like it's producing cherries instead of berries.

Among the 2,000 trees and shrubs Bill Cannon has on his one-acre farm in Brewster are more than 320 cultivars of holly. His father, a florist, was a holly fancier before him. One of the bushes the senior Cannon planted 80 years ago is now 35 feet tall and the largest variegated holly in New England.

Like father, like son. Over the years, Cannon has added a tremendous number of unusual hollies to the original collection and now the collection is the largest in New England. Each plant is neatly identified with its own tag, which he verifies when he prunes every year.

A former Holly Society of America president, garden writer, horticulturist, and graduate of the Stockbridge School at UMass Dartmouth, he is knowledgeable and curious about plants in general. His greenhouse is constantly filled with specimens he is experimenting with or propagating. Cannon admires and collects crepe myrtle and magnolias, but he is hooked on holly.

An expert at cross-pollination, he has three registered and named holly cultivars of his own: Kelsey's Delight, Gretchen, and Purple Frost. According to him, holly grows easily on the Cape and hundreds of different cultivars thrive in the maritime climate. Most people have no idea of the variety available. Unlike the popular type seen at Christmas, some have orange, yellow, or black berries, and the leaves don't resemble holly at all. He finds it all fascinating, but also enjoys the plants for other reasons.

The plants offer winter shelter to birds and other wild creatures, and the berries provide food. Holly shrubs also make an excellent hedge and privacy screen. Once a plant is established it also makes a good windbreak. Most hollies

Brewster resident Bill Cannon, former president of the Holly Society of America, grows a lot of different plants in his greenhouse, but his specialty is holly.

A visit to the holly farm provides insight into the huge variety of hollies available.

Cannon created an allée—a grass walkway—in his rear garden with specimen hollies on each side, interplanted with evergreen magnolias and crepe myrtle.

CLOCKWISE:

Each bush is pruned, watered, nurtured and tagged.

Ilex attenuata "Alagold" surprises with gold berries.

Ilex cornuta, with its unusual clustered berries, is also known as "Berries Jubilee."

If you have holly and want to know if the plants are male or female, go out in the spring and inspect them. The females will have flowers with a berry center.

119

are evergreen and add color to the winter landscape. They fit in well with pine and fir trees as well as rhododendrons and azaleas. And there is a bonus. There are no needles or leaves to clean up in the fall.

He recommends blue hollies for the novice gardener: Blue Prince, Blue Stallion, and Blue Boy. Honey Maid is a another hardy holly that is variegated, or try the Koni hollies.

In November and December, Cannon sells cut-greens to local florists, nurseries, neighbors, and garden clubs and also takes orders for custom-made wreaths. In heraldry, holly symbolized truth, which may be why it is still used in holiday decorations today.

Cannon is a horticulturist who loves all plants—just holly more than others. He propagates a variety of plants in his greenhouse.

Aquatic plants winter over in the comfort of the greenhouse.

Tips from the top

Most hollies are well adapted to the Cape.

Holly is a fairly carefree plant with few problems.

They are shallow-rooted and benefit from being mulched. It keeps the soil from changing temperature too quickly and builds up the soil at the same time. Dig the hole wide rather than deep. Mix in peat moss or compost and be sure there is good drainage.

Slow release fertilizer can be added right over the mulch in April and June. Don't stimulate growth in the fall.

Holly likes a full or half day of sun. It gets leggy in the shade.

Holly can be pruned at any time so do it during the holiday season and use the greens for decorations.

Both male and female hollies are needed for cross pollination.

Hollies bloom on old wood.

Deer and rabbits will nibble on new growth so it needs to be protected. Use a repellent spray or build a barrier with chicken wire and stakes around the plant to discourage them.

Holly Farm, Route 6A, Brewster. Plants for sale. Holly and boxwood wreaths in season by special order.

FOR MORE INFORMATION

call: 508-896-3740
email: ilexbc@verizon.net

THACHER
CRANBERRIES
HARWICH

Raymond "Linc" Thacher and his son, Ray, Jr., a 13th generation Thacher, have worked the bogs together since Ray was old enough to pick.

At the height of his career, Linc Thacher was harvesting 350 acres of cranberries a year. These days, he's cut it back to 60. Over the last half century, he's seen prices soar to $66 a barrel and bottom out at $12. In fact, he's seen about everything that has to do with cranberries, from pollinator problems to frozen bogs.

Thacher bought his first bog in Brewster in 1953 and says every year is unpredictable. But Thacher has one thing he counts on. Family. His son Ray and daughter Jayne have helped with the harvest since they were eight years old. Now his grandson Caleb also joins him in the bogs and screening room. The grandfather has handed the reins to his son, Ray, but he is still around to make sure he holds them tight. And he would never miss a harvest.

Men may have landed on the moon but cranberries are still picked the old fashioned way—manually. When it isn't picking time, there is plenty of preparation, maintenance, weeding, and upkeep of equipment to keep a grower busy. And it is harder to find old-fashioned labor.

LEFT TO RIGHT:
A lightweight vehicle is used to pick up the sacks so the vines aren't crushed.

The dry-picked berries go into burlap sacks and are left in the bog for pickup.

The best berries are dry-picked first, the old-fashioned way, and command a higher price on the market as fresh fruit.

INSET, TOP TO BOTTOM:
Leo Cakounes shares his cranberry crumble with a friend. He works for the Thachers and they reciprocate by helping him on his bog.

Workers take a break and enjoy the quiet of the bog.

The Thachers are one of 23 farmers who grow for Ocean Spray on the Cape. In winter, they flood the bogs to protect tender vines from freezing. It's also the time of year that sand is added, which is needed to help root the cranberry plants. In the spring, insect-eating birds like swallows are welcomed with birdhouses on the bogs. It's a natural way to help control unwanted worms and bugs that could destroy the crop. Sprinkler systems must be at the ready in case of a sudden frost.

When the warm days of June arrive, the plants begin to flower and the Thachers rent bee hives to ensure pollination. Meanwhile, the banks around the bogs are mowed to keep grass, poison ivy, and weeds under control. Once the berries form and begin to grow, they must be irrigated regularly.

The berries ripen in late September and October. The best are dry-picked first as fresh fruit and they command a higher price. Later, the rest are wet-picked in flooded bogs. Ray drives a thrasher to shake them free from the vines. They float until they are rounded up with boards and corralled to a machine where they are sucked up a tube and into a waiting truck. Wet-picked berries are made into juice and processed products. The grower is paid by the barrel (100 pounds) even though the berries are actually loaded into plastic bins these days. An acre can produce 100 barrels.

The Thachers' work is not over with a trip to the bank, however. All those birdhouses have to come down, the sprinkler system dismantled, and the harvesting equipment prepared for winter. Then, it's time to start sanding and flooding and preparing for spring, and the cycle begins again.

COUNTER CLOCKWISE:
Workers corral the floating berries into a circle.

Some bogs are better suited to wet picking. Fortunately, berries float.

Ray Thacher has a bog right outside his front door. In December he puts a lighted Christmas tree in the middle for the neighborhood to enjoy.

COUNTER CLOCKWISE:
Workers put the final touches on a circle of berries.

Ray Thacher loosens the berries from the vines.

The berries are vacuumed out of the water and into the waiting truck that will take them
to Ocean Spray, where they will be made into juice.

127

LEFT TO RIGHT:
Dry-picked berries are put through the separator and then sold to local markets.

Workers will return for the rest of the berries tomorrow.

OPPOSITE
Cranberry farming is done during the beautiful days of autumn on the Cape.

Fresh berries freeze well and can be used throughout the year
in many recipes

Tips from the top

Classic cranberry sauce is made from 1 cup sugar, 1 cup water, and 4 cups fresh cranberries (orange peel optional). Wash cranberries. Heat water and sugar to a boil and stir to dissolve sugar. Add cranberries, return to a boil, reduce heat, and simmer for 10 minutes until cranberries pop. Cool completely at room temperature and then chill in refrigerator. Sauce thickens as it cools. Recipe makes 2 1/4cups.

Raymond L. Thacher and Son, Cranberries at 359 Main St., Harwich, sells cranberries in season at local farmers markets and grocery stores.

FOR MORE INFORMATION

call: 508-432-2931

TRURO
AGRICULTURAL
FAIR
TRURO CENTER

Alissa Dewitt works on the vegetable beauty contest.

Imagine a day filled with music, homemade pies, colorful produce, fresh flowers, clams, oysters, herbs, fish, honey, pony rides, and sack races. Francie Randolph did. And she made it happen. The first Truro Agricultural Fair was held in 2009 and has grown exponentially, thanks to Randolph and the co-founders, volunteer director Stephanie Rein, and participation director David Dewitt.

Francie and her husband, Thomas A. D. Watson, grow most of their own food. Both artists, they live a bucolic life on a three-acre farm with their two children in a 200 year-old house. They also tend bees, chickens, ducks, sheep, and a hog. Thomas fishes and preserves his catch in his own smokehouse. Their lifestyle is sustainable and they teach by example. What they don't consume they freeze for the winter months.

LEFT TO RIGHT:
Francie Randolph and her husband Thomas A. D. Watson grow their own food on a three-acre farm with bees, chickens, ducks, sheep, and a hog. Their lifestyle is sustainable and they teach by example.

Francie Randolph came up with the idea of a local agricultural fair for Truro and made it happen with the help of co-founders, volunteer director Stephanie Rein and participation director David Dewitt.

LEFT TO RIGHT:
Each year more and more people come to this old-fashioned agricultural fair, which showcases a simple, wholesome way of life.

Vegetables are artfully displayed and farmers are happy to give tips on growing them.

Despite her degree from Harvard and his from the Rhode Island School of Design, they live by their hands, whether painting or planting.

Having had fond memories of the West Tisbury Farmers Market on Martha's Vineyard where she spent summers, Francie conceived of a similar fair for Truro. She wanted to showcase the local agricultural community and the simple, wholesome way of life she cherishes. With the help of neighbors Stephanie Rein and David Dewitt, they got in touch with farmers, rounded up 50 volunteers, and hung 100 posters all over town.

More than 2,500 people showed up the first year. It has continued to double in size every year since.

The success has to do with the charm of the fair. It's a Norman Rockwell world with pie-eating contests, live bluegrass music, and farmers showing off their wares. There is even a beauty contest—for animals and vegetables. Farmers provide educational material as well; it's not just about selling fresh produce but sharing knowledge. It's like stepping back a century in time when you see wool being spun and children in a burlap sack race. It makes you wish there was a fair every weekend.

The mission statement is: "The Truro Agriculture Fair promotes a broader appreciation of the agriculture, aquaculture, fishing, and farming taking place on Cape Cod today. It showcases our agrarian history, the foods currently grown and harvested, and connects the health of our local food system to our environment and future."

COUNTER CLOCKWISE:
Children take to the races.

A pie-eating contest creates a Norman Rockwell moment and a lot of sticky faces.

Alpaca make nice pets and their fleece can be spun into wool.

The village green is the perfect setting for the fair.

Morgan Seaman of Dover selects her favorite pumpkin.

Truro Agricultural Fair, 7 Truro Center Road, Truro Center, is usually held the Sunday of Labor Day weekend in September, from 10 a.m. to 4 p.m.

FOR MORE INFORMATION

website: www.truroagfair.com

SATUCKET
FARM STAND
BREWSTER

Anderson grows flowers for the farm stand in her garden.

Anita Anderson and her husband, Cy Rourke, have operated this farm stand since 1996, but the open air market has been around for over a 100 years. It's a neighborhood place that hires local teenagers to work at the stand and keep Annie, a big, friendly yellow lab, company. Anderson finds local bakers to supply the stand with pies, muffins, breads, and pastries, and makes sure they use all natural ingredients with no preservatives. The same is true of the jams, jellies, and the local cheeses. The produce is native as well, and includes corn, tomatoes, organic strawberries, raspberries, and blueberries, all in season. It is the epitome of the Buy Fresh, Buy Local movement, so it is no wonder they have won fifteen consecutive awards from a local magazine. Anita is a gardener in addition to being a businesswoman, and maintains the beautiful grounds that supply the business with freshly arranged bouquets and seasonal herbs.

LEFT TO RIGHT:
Anderson even plants the basket of the old bike in front of the farm stand.

Anita Anderson has operated the open air Satucket Farm Stand since 1996, but it's been around for over a century.

CLOCKWISE:

It's a wonderful life when there are such great choices of jams, jellies, and culinary treats.

Local bakers are sought to provide fresh delicacies for the stand.

Local tomatoes are available in all sizes.

The local farm stand is picturesque and a fun spot to visit for local goods.

139

Satucket Farm Stand, 76 Harwich Road, Route 124, Brewster, is open Memorial Day to Labor Day. Hours are 6:30 a.m. to 6:30 p.m.

FOR MORE INFORMATION

call: 508-896-5540
email: satucketfarm@comcast.net
website: www.satucketfarm.com

TOP TO BOTTOM:
A spider web outlines the morning's dew.

Ferns make good companions for flower bouquets.

TONY ANDREWS
FARM
EAST FALMOUTH

Kaydence Storm, with her mother Michelle, makes her choices for Halloween.

The Tony Andrews Farm has been well-known in the Falmouth area since 1935, when it began servicing the community with fresh produce, straw, hay bales, fruit, and flowers. It is family friendly and offers some pick-your-own options and hayrides.

LEFT TO RIGHT:
The Tony Andrews Farm specializes in pick-your-own pumpkins, with lots of wide open fields to make a selection.

Choose the perfect pumpkin at the farm stand or wander through the pumpkin patch.

TOP TO BOTTOM:

Indian corn makes a fine seasonal decoration in autumn.

Gourds come in a variety of colors and shapes.

Pick a bouquet of sunflowers to take home from the fields.

The farm is located at 394 Old Meetinghouse Road, East Falmouth, and is open daily from 10 a.m. to 6 p.m. with fresh fruit, native corn, and vegetables. Pick your own strawberries, peas, sunflowers, and pumpkins in season. A picnic area, fall decorations, and hay rides are also available. School groups are welcome by reservation.

FOR MORE INFORMATION

call: 508-548-4717
email: tonyandrewsfarm@comcast.net